Junior Drug Awareness

Inhalants

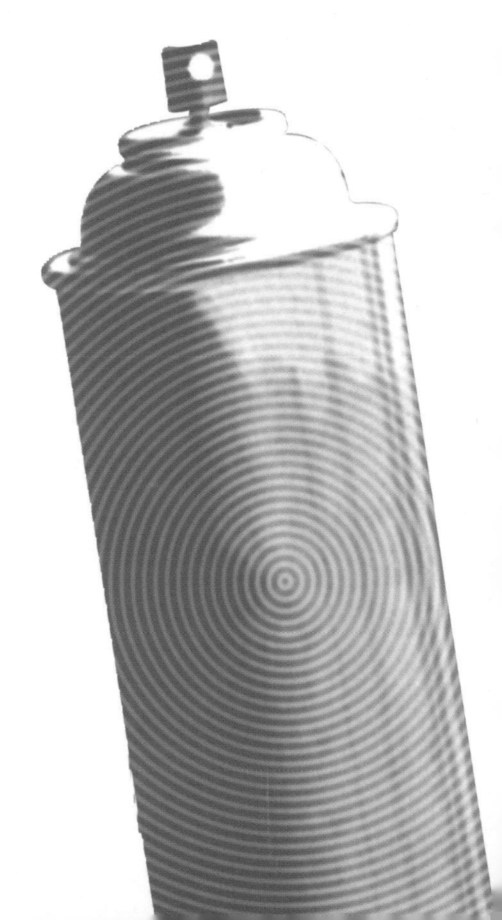

Junior Drug Awareness

Alcohol

Amphetamines and Other Uppers

Crack and Cocaine

Ecstasy and Other Designer Drugs

Heroin

How to Get Help

How to Say No

Inhalants and Solvents

LSD, PCP, and Other Hallucinogens

Marijuana

Nicotine and Cigarettes

Pain Relievers, Diet Pills, and
Other Over-the-Counter Drugs

Prozac and Other Antidepressants

Steroids

Valium and Other Downers

Junior Drug Awareness

Inhalants

Introduction by **BARRY R. McCAFFREY**
Director, Office of National Drug Control Policy

Foreword by **STEVEN L. JAFFE, M.D.**
Senior Consulting Editor, Professor of Child and Adolescent Psychiatry, Emory University

Linda N. Bayer, Ph.D.

Chelsea House Publishers
Philadelphia

The author wishes to dedicate this book to her son, Philip Lev Bayer, whose courage, sensitivity, wisdom, and compassion far surpass the depth that normally marks someone of his years.

In Hebrew, lev means "heart." Lev is all heart; he is a comfort and inspiration to us all.

For young people torn by substance abuse and other painful problems: take heart. Help is at hand.

CHELSEA HOUSE PUBLISHERS
Editor in Chief Stephen Reginald
Production Manager Pamela Loos
Director of Photography Judy L. Hasday
Art Director Sara Davis
Managing Editor James D. Gallagher
Senior Production Editor Lisa Chippendale

Staff for INHALANTS
Senior Editor Therese De Angelis
Associate Art Director/Designer Takeshi Takahashi
Picture Researcher Patricia Burns
Cover Illustration/Design Keith Trego

http://www.chelseahouse.com

3 5 7 9 8 6 4

Library of Congress Cataloging-in-Publication Data
Bayer, Linda, 1948-
Inhalants and solvents / by Linda Bayer.
80 pp. cm. — (Junior drug awareness)
Includes bibliographical references and index.
Summary: Discusses how and why people abuse inhalants and solvents, the dangerous effects these substances can have on the body, and where to get help.
ISBN: 0-7910-5178-1 (hardcover)
1. Solvent abuse—Juvenile literature. 2. Solvents—Health aspects—Juvenile literature. [1. Solvent abuse. 2. Substance abuse.] I. Title. II. Series.
HV5822.S65B39 1999
362.29'9—dc21 98-49788
CIP
AC

Contents

INTRODUCTION

by Barry R. McCaffrey
Director, Office of National Drug Control Policy

STAYING AWAY FROM ILLEGAL DRUGS, TOBACCO PRODUCTS, AND ALCOHOL

Good health allows you to be as strong, happy, smart, and skillful as you can possibly be. The worst thing about illegal drugs is that they damage people from the inside. Our bodies and minds are wonderful, complicated systems that run like finely tuned machines when we take care of ourselves.

Doctors prescribe legal drugs, called medicines, to heal us when we become sick, but dangerous chemicals that are not recommended by doctors, nurses, or pharmacists are called illegal drugs. These drugs cannot be bought in stores because they harm different organs of the body, causing illness or even death. Illegal drugs, such as marijuana, cocaine or "crack," heroin, methamphetamine ("meth"), and other dangerous substances are against the law because they affect our ability to think, work, play, sleep, or eat.

If anyone ever offers you illegal drugs or any kind of pills, liquids, substances to smoke, or shots to inject into your body, tell them you're not interested. You should report drug pushers—people who distribute these poisons—to parents, teachers, police, coaches, clergy, or other adults whom you trust.

Cigarettes and alcohol are also illegal for youngsters. Tobacco products and drinks like wine, beer, and liquor are particularly harmful for children and teenagers because their bodies, especially their nervous systems, are still developing. For this reason, young people are more likely to be hurt by illicit drugs—including cigarettes and alcohol. These two products kill more people—from cancer, and automobile accidents caused by intoxicated drivers—than all other drugs combined. We say about drug use: "Users are losers." Be a winner and stay away from illegal drugs, tobacco products, and alcoholic beverages.

Here are four reasons why you shouldn't use illegal drugs:

- Illegal drugs can cause brain damage.
- Illegal drugs are "psychoactive." This means that they change your personality or the way you feel. They also impair your judgment. While under the influence of drugs, you are more likely to endanger your life or someone else's. You will also be less able to protect yourself from danger.
- Many illegal drugs are addictive, which means that once a person starts taking them, stopping is extremely difficult. An addict's body craves the drug and becomes dependent upon it. The illegal drug–user may become sick if the drug is discontinued and so may become a slave to drugs.

- Some drugs, called "gateway" substances, can lead a person to take more-dangerous drugs. For example, a 12-year-old who smokes marijuana is 79 times more likely to have an addiction problem later in life than a child who never tries marijuana.

Here are some reasons why you shouldn't drink alcoholic beverages, including beer and wine coolers:

- Alcohol is the second leading cause of death in our country. More than 100,000 people die every year because of drinking.
- Adolescents are twice as likely as adults to be involved in fatal alcohol-related car crashes.
- Half of all assaults against girls or women involve alcohol.
- Drinking is illegal if you are under the age of 21. You could be arrested for this crime.

Here are three reasons why you shouldn't smoke cigarettes:

- Nicotine is highly addictive. Once you start smoking, it is very hard to stop, and smoking cigarettes causes lung cancer and other diseases. Tobacco- and nicotine-related diseases kill more than 400,000 people every year.
- Each day, 3,000 kids begin smoking. One-third of these youngsters will probably have their lives shortened because of tobacco use.
- Children who smoke cigarettes are almost six times more likely to use other illegal drugs than kids who don't smoke.

If your parents haven't told you how they feel about the dangers of illegal drugs, ask them. One of every 10 kids aged 12 to 17 are using illegal drugs. They do not understand the risks they are taking with their health and their lives. However, the vast majority of young people in America are smart enough to figure out that drugs, cigarettes, and alcohol could rob them of their future. Be your body's best friend: guard your mental and physical health by staying away from drugs.

FOREWORD

WHY SHOULD I LEARN ABOUT DRUGS?

Steven L. Jaffe, M.D., Senior Consulting Editor,
Professor of Child and Adolescent Psychiatry,
Emory University

Your grandparents and great-grandparents did not think much about "drug awareness." That's because drugs, to most of them, simply meant "medicine."

Of the three types of drugs, medicine is the good type. Medicines such as penicillin and aspirin promote healing and help sick people get better.

Another type of drug is obviously bad for you because it is poison. Then there are the kind of drugs that fool you, such as marijuana and LSD. They make you feel good, but they harm your body and brain.

Our great crisis today is that this third category of drugs has become widely abused. Drugs of abuse are everywhere, not just in rough neighborhoods. Many teens are introduced to drugs by older brothers, sisters, friends, or even friends' parents. Some people may use only a little bit of a drug, but others who inherited a tendency to become addicted may move on to using drugs all the time. If a family member is or was an alcoholic or an addict, a young person is at greater risk of becoming one.

Drug abuse can weaken us physically. Worse, it can cause permanent mental damage. Our brain is the most important part of our body. Our thoughts, hopes, wishes, feelings, and memories are located there, within 100 billion nerve cells. Alcohol and drugs that are abused will harm—and even destroy—these cells. During the teen years, your brain continues to develop and grow, but drugs and alcohol can impair this growth.

I treat all types of teenagers at my hospital programs and in my office. Many suffer from depression or anxiety. A lot of them abuse drugs and alcohol, and this makes their depression or fears worse. I have celebrated birthdays and high school graduations with many of my patients. But I have also been to sad funerals for others who have died from problems with drug abuse.

Doctors understand more about drugs today than ever before. We've learned that some substances (even some foods) that we once thought were harmless can actually cause health problems. And for some people, medicines that help relieve one symptom might cause problems in other ways. This is because each person's body chemistry and immune system are different.

For all of these reasons, drug awareness is important for everyone. We need to learn which drugs to avoid or question—not only the destructive, illegal drugs we hear so much about in the news, but also ordinary medicines we buy at the supermarket or pharmacy. We need to understand that even "good" drugs can hurt us if they are not used correctly. If we take any drug without a doctor's advice, we are taking a risk.

Drug awareness enables you to make good decisions. It allows you to become powerful and strong and have a meaningful life!

A homeless boy in the city of Manila, the Philippines, sniffs solvent from a bottle. People from low-income regions or families are not the only ones who abuse solvents and inhalants. Since most of these substances are cheap and easy to find, young people aged 12 to 17 are the most frequent abusers of these dangerous chemicals.

1 COMMON BUT DANGEROUS SUBSTANCES

A 13-year-old boy inhaled fumes from cleaning fluid and became ill within a few minutes. The boy's friends told his parents, who took him to the hospital. He was put on life-support systems, yet his condition did not improve. The doctors did everything they could to help the boy, but he died within 24 hours.

An 11-year-old girl collapsed in a public bathroom. A container of butane (cigarette lighter fluid) and a plastic bag were on the floor next to her. Bottles of correction fluid were found in her pockets. Other people in the bathroom tried to revive the girl, but she never regained consciousness. She was pronounced dead soon after.

A 15-year-old boy fainted suddenly in his backyard. According to three other kids who were there, the four had taken gas from the propane tank that was attached to the family's outdoor grill. They put the gas in a bag and inhaled it to try to get high. The boy who fainted died on the way to the hospital.

Maybe you already know what inhalants are. But do you know what makes them dangerous? When a person inhales fumes, whether deliberately or accidentally, from any of these products, the chemicals enter the lungs and brain. The person becomes dizzy and may also feel nauseated, experience blurred vision, or have trouble remembering things. Inhalants are also **psychoactive** (mind-altering), meaning that they can change the way the user thinks and behaves. He or she may even lose consciousness or die.

Getting the Facts Straight

Some people don't realize that common products sold in stores are actually drugs that can be abused by inhaling them or their vapors. People may think that drugs are illegal substances like marijuana ("pot"), heroin, or cocaine without realizing that inhalants—as well as cigarettes and alcohol—are also harmful drugs. Most substances that people consume that aren't food (and therefore have no nutritional value) are drugs.

Some drugs are prescribed by doctors to treat illness; others, like cold remedies and pain relievers, are sold in drugstores and supermarkets. With the exception of alcohol and tobacco, we call these drugs medicines. However, if a drug has no medical use and can cause injury or death, it is usually illegal to buy, sell, or use.

One difference between inhalants and drugs like marijuana, heroin, and cocaine is that inhaled sub-

Amyl nitrite (shown here being used as a "popper" to get high) is one of the few substances abused as an inhalant that is controlled and dispensed by prescription only. Doctors prescribe it for some patients to ease the pain caused by restricted blood flow to the heart.

stances are usually legal. That's because they are sold for purposes other than abusing them to get high. When used properly (according to the directions on the package), solvents are fairly safe. But when people try to use these products in other ways, such as inhaling the products' fumes, they are risking their lives. Solvents damage

ABUSABLE PRODUCTS

Volatile Organic Solvents

- Adhesives: model airplane glue, rubber cement, household glue
- Aerosols: spray paint, hair spray, air freshener, deodorant, fabric protector
- Solvents and Gases (products that are abused in liquid or gas form): nail polish remover, paint thinner, correction fluid and thinner, toxic markers, toluene, cigarette lighter fluid, gasoline, octane booster
- Cleaning Agents: dry-cleaning fluid, spot remover, degreaser
- Food Products: vegetable cooking spray, dessert topping spray

Anesthetics

- Chloroform, ether, nitrous oxide

Nitrites

- Amyl nitrite, butyl nitrite (also marketed as video head cleaner), nitrite room odorizers (also called "poppers," "rush," or "locker room")

Source: National Inhalant Prevention Coalition, 1998

All inhalants are gases or solvents of some type. The National Inhalant Prevention Coalition (NIPC) categorizes these substances according to their original or intended uses. Here is the NIPC's list. Do you know anyone who has abused these substances?

the brain and other organs, and they can be just as dangerous as illegal drugs like cocaine and heroin—sometimes even more dangerous. Along with alcohol, tobacco products, and marijuana, inhalants are among the substances most likely to be abused by young people.

Here are some everyday household products that can be classified as inhalants:

- butane (cigarette lighter fluid)
- correction fluid (such as Wite-Out and Liquid Paper)
- felt-tip markers
- furniture polish
- gasoline
- gas used in whipped cream cans
- hair spray
- model airplane glue
- nail polish remover
- shoe polish
- spray paint
- turpentine (paint thinner)

Types of Inhalants

We can place these substances and others like them into two categories: gases and **volatile organic solvents**. Most inhalants are part of the second group.

What exactly is a volatile organic solvent? Let's look at the meaning of each word in that phrase. A **solvent** is a liquid (such as water, alcohol, or ether) that can dissolve another substance to form a solution. A **volatile**

All of the common household and school products shown here—felt-tip markers, nail polish remover, rubber cement, and nail polish—are volatile organic solvents.

substance is a liquid that rapidly changes into a vapor. The word "volatile" also describes liquids that can easily explode or catch fire, such as paint thinner, model airplane glue, and gasoline. An **organic** substance is one that contains two of nature's basic elements: carbon and hydrogen. One of the reasons solvents can interfere with the body's functions is that all living organisms—you, your dog, a rose bush—also contain carbon and hydrogen and are therefore organic.

Did you know that beers, wines, and liquors are solvents? The first known case of solvent abuse was **ethanol** drinking. Ethanol is the type of alcohol found in beer, wine, and hard liquors. It is produced by a chemical process called **fermentation**, in which yeast consumes the sugars in fruits or grains and produces the alcohol and a gas. The higher the level of sugar, the greater the alcohol concentration. But even when the alcohol concentration is increased by another process called distilling (boiling off the alcohol and collecting it), the ethanol is not volatile enough to produce strong effects when sniffed. That's why people drink ethanol rather than inhale its vapors.

Some inhalants belong to another category: gases. An example is nitrous oxide, which is sometimes called "laughing gas" because it makes people feel giddy. Because it also has **anesthetic** qualities (meaning it relieves pain due to a loss of feeling or consciousness), nitrous oxide is often used by dentists to ease the discomfort of dental surgery.

What Inhalants Can Do to You

When a drug is taken through the mouth (like a pill), it enters the bloodstream slowly, because it has to pass through the digestive system just like food does. The drug takes longer to affect the brain, and the impact on the body is usually less intense than if it were injected or breathed in. Drugs that are injected or inhaled bypass the stomach and intestines and reach the brain very quickly. For this reason, most drugs are more harmful when they are injected, **"sniffed"** (inhaled through the nose), or **"huffed"** (inhaled through the mouth) than when they are swallowed.

Volatile organic solvents act as **depressants**, which means that at first they slow down your central nervous system the way alcohol does. But they are also **toxic** (poisonous) and can cause blurred vision, dizziness, **tremors** (uncontrollable shaking), slurred speech, and drowsiness. They can damage your liver, kidneys, and other organs. Some can cause blindness or hearing loss. They can easily stop your heart and kill you.

Volatile organic solvents also disrupt the **membranes** (thin walls) of nerve cells and interrupt impulses to and from the brain. When this happens, a kind of brain paralysis occurs that can be fatal. The more these solvents are inhaled, the more damage they cause.

Gases like nitrous oxide, amyl nitrite (used by heart patients), and butyl nitrite cause people to feel lightheaded for a few minutes. Abusing gases like these increases a person's heart rate. These gases can also cause the heart to beat irregularly. A person who abuses whippets (inhaling the gas in whipped cream containers through the mouth) may faint or go into a **coma** (a deep state of unconsciousness from which the person might not awaken). Inhalants can kill you, even the first time you use them.

We have already learned about another great danger with inhaled substances. They are volatile—they can easily catch fire or explode while being used, especially if they are exposed to high heat or flames. And because

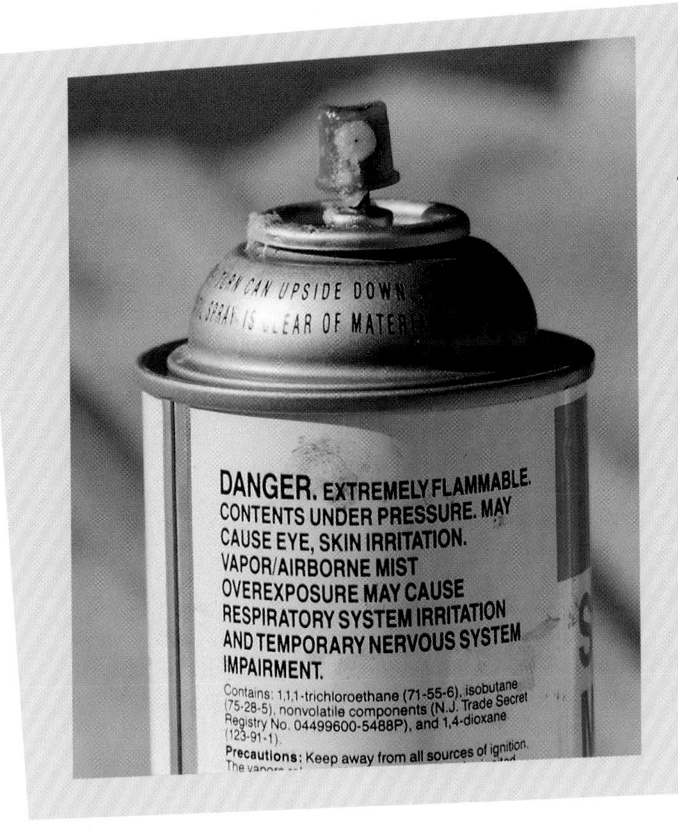

Even if you don't intend to abuse products like cleaning fluids, office supplies, or art materials by inhaling their fumes, you must still be careful while handling them. It's important to read warning labels, such as the one found on this spray adhesive can, so that you know how to handle these products safely.

they can reduce the ability to think clearly, a user is unlikely to pay attention to the increased danger. Some inhalant abusers have been badly burned or killed when **flammable** (easily ignitable) solvents caught fire.

Why Use Inhalants?

If inhalants are so dangerous, why do some people abuse them? Most kids—four out of five, in fact—never do. However, one in five young people abuse inhalants some time before graduating from high school. Youngsters try inhalants as a quick way to feel "high" or intoxicated. Inhalants are usually cheaper and easier to get than illegal drugs, which makes some kids think that they're not as dangerous.

But make no mistake: inhalants are drugs. Taking poisons into the body through your nose or mouth injures your lungs, heart, kidneys, liver, and other organs. It can also cause permanent brain damage or instant death. Youngsters who inhale toxic chemicals are endangering their health and their lives.

Even if you don't abuse inhalants by deliberately breathing them in, the fumes from cleaning products, office supplies, and art materials like those listed earlier in this chapter can still be harmful. Always read the instructions on these products or ask an adult to be present while you're using them. Make sure you are handling them properly and safely.

All of the tragedies you read about at the beginning of this chapter might have been prevented if the kids involved had used common sense, or if their friends or family members had discovered what was going on and stopped them. Sometimes children—and even adults—can do silly or dangerous things because they think others will laugh at them or think less of them if they don't. When you feel as though you need to follow the behavior of members of a group instead of following your own conscience, you are experiencing **peer pressure**. Imagine how bad the friends of these kids felt later when they realized that they could have saved their friends' lives by telling parents, teachers, or other adults about the inhalants.

If you suspect that someone you know is abusing solvents by sniffing or huffing, don't be afraid to tell your parents, guidance counselor, coach, school nurse, or another adult you trust. Try not to feel bad about seeking help. After all, good friends care about their friends and want them to be healthy and safe. Getting help may be the best thing you can do for someone in trouble. In Chapter 5, we'll look at some of the ways you can tell whether a friend or family member has been abusing inhalants or other drugs and how you may be able to help.

It's important that kids like you learn how risky drug use can be, so that you can make an informed decision to stay drug-free. September 4 has been designated National Inhalant Abuse Prevention Day. By learning more about the dangers of inhalants and solvents, you can take a big step toward protecting your health and that of your friends and family.

This Native American medicine man from the early 1800s may have inhaled the smoke from burning plants or used mind-altering substances as part of his healing ritual.

2 A HISTORY OF INHALANTS AND SOLVENTS

You may be surprised to learn that inhalant and solvent abuse is not new. Throughout recorded history, people have inhaled substances to experience powerful effects. At the oracle (or prophet) of Delphi in ancient Greece, a priestess named Pythoness sat over a fire of burning laurel leaves and experienced visions when she inhaled the carbon dioxide that was produced. The Greeks believed that when the priestess was in this state, she acted as the spokesperson for their gods.

In biblical Palestine and in ancient Egypt, ointments and perfumes were used to enhance religious worship. Archaeologists have discovered stone altars from the ancient city of Babylon and from Palestine that were used to burn incense made from aromatic (fragrant) woods and spices. In many cases these seemingly harmless substances were actually psychoactive drugs. In the

When nitrous oxide was first discovered, people were unaware of its dangerous side effects. It was so popular that people held parties specifically to get high on the drug. Today, however, it is used only by dentists to relieve pain for some of their patients.

Mediterranean islands 2,500 years ago and in Africa hundreds of years ago, marijuana leaves and flowers were thrown into fires, and the smoke was inhaled. During religious ceremonies, North and South American Indians often inhaled or ingested **hallucinogenic** substances.

Actually, many substances, such as tobacco and opium, will produce unusual effects when they are smoked or inhaled. Scientists now know that the state of mind produced by inhaling these substances is partly the result of **anoxia**, or a lack of oxygen. Humans need oxygen to breathe properly, and the combination of holding their breath and absorbing carbon dioxide from the smoke creates a state of anoxia in the users. The degree to which anoxia contributes to the effects of these drugs is unknown, but we now know that at least part of the euphoria that is experienced is due to oxygen deprivation.

Inhalants as Pain Relievers

As more volatile agents were produced, people began to notice that inhaling these agents caused profound behavioral changes. In 1831, a chemical called **chloroform** was discovered in three different parts of the world—Germany, France, and the United States. Before long, chloroform was used as an anesthetic during surgery and childbirth.

Chloroform became so popular that Queen Victoria of England received it during the delivery of her eighth child. Samuel Guthrie, the American researcher who discovered chloroform, wrote a paper that described the effects of the substance and declared that there seemed to be little danger in using it: "A great number of persons have drunk of the solution . . . in my laboratory," he wrote, "and so far as I have observed, it has appeared to be singularly [agreeable], both to the palate [its taste] and stomach, producing a lively flow of animal spirits." Guthrie also reported that chloroform was better than alcohol because it did not leave the user

In the 1990s, inhalant abuse was more common among younger teens than among high school kids. This chart, prepared by the National Institute on Drug Abuse, shows that lifetime use of inhalants ranked third—after alcohol and cigarettes—among eighth graders in 1993. During that same period, inhalants ranked fourth (after alcohol, cigarettes, and marijuana) among tenth and twelfth graders.

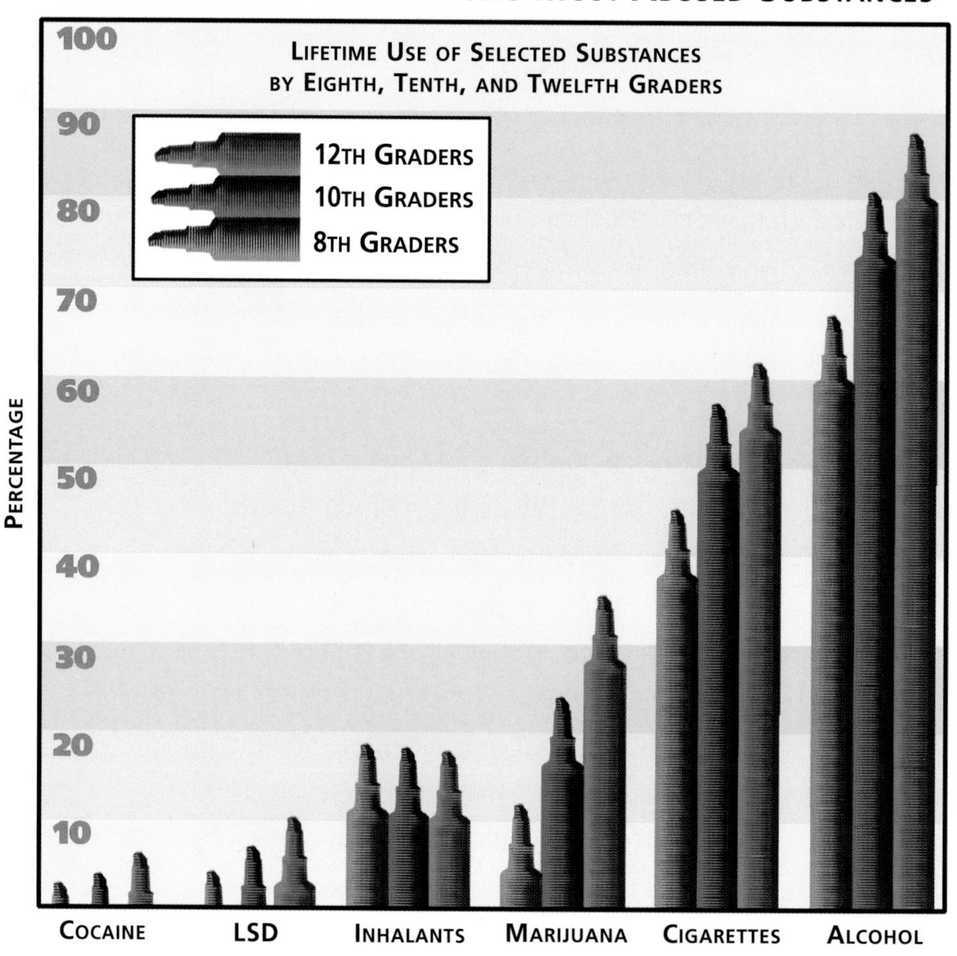

Source: NIDA Monitoring the Future Study, 1993

depressed after its euphoric effects wore off.

The same year chloroform was discovered, however, overdoses of the substance became a serious problem. The first person in the country to use nitrous oxide for a surgical procedure, American dentist Horace Wells, died after abusing chloroform. Eventually, other anesthetics became more popular, and chloroform abuse subsided.

Nitrous oxide had been discovered in 1776 by Sir Joseph Priestly. The drug became so popular that people held nitrous oxide parties and demonstrations where guests could experience the lightheaded euphoria (a feeling of well-being) and uncontrollable fits of laughter that came with inhaling the drug. In 1799 Sir Humphry Davy, the man who first synthesized the drug, observed that it seemed to have pain-reducing qualities and should be considered for use during surgery. But it was not tested for this purpose until 1844.

There are reports of American students using nitrous oxide for recreational purposes in the early 19th century. One student, realizing he could make a great deal of money, actually quit medical school and went into the nitrous oxide business. He held demonstrations and sold the inhalant for 25 cents per dose. At one of these gatherings, Horace Wells watched an intoxicated man seriously injure himself without feeling pain. The following day he used the inhalant while the student pulled one of his teeth—and he realized that nitrous oxide could change dentistry for good.

Another inhalant with anesthetic properties is **ether**. In the late 1800s, ether was widely used as a medicine.

Unlike nitrous oxide or chloroform, however, it was often consumed as a liquid. As with nitrous oxide, some people began holding "ether frolics" or parties, during which they would drink ether as if it were a liqueur.

Inhalant Abuse

Between the mid-1800s and the mid-1900s, many cases of solvent abuse were reported. In most of these cases, the abusers had first been given the drug to treat a medical problem. One doctor prescribed a painkiller and anesthetic called trichloroethylene to treat a man's facial twitch, but the man continued taking the drug for its mind-altering effects. Another patient began inhaling chloroform to relieve headaches but developed a habit of using the drug regularly. The amount he used each day increased until he consumed an entire bottle and was unconscious for four days.

In these examples, the two patients developed what is called **compulsive** behavior—a strong urge that is very difficult to control. Although the use of volatile organic solvents was well known by the beginning of the 20th century, scientists and physicians were not yet aware of the compulsive nature of this practice and did not recognize it as a form of drug abuse. But the problem began to attract public attention in the 1940s and 1950s. The earliest accounts came from the East Coast. One such story describes an outbreak of gasoline sniffing in Warren, Pennsylvania, in the late 1940s. Newspapers and magazines began to focus on the increasing number of youngsters sniffing model airplane glue. As a result,

All of the products shown here—hair spray, air freshener, disinfectant, and cooking spray—are aerosols and are categorized as volatile organic solvents.

toluene, the mind-altering solvent in the glue, was later removed from glue and many spray paints.

In the 1960s and 1970s, inhalants, like many other mind-altering drugs, became popular among young people. The first case of glue sniffing in Great Britain was reported in 1962. After 18 months, a 20-year-old man had increased his dose from one-third of a tube of glue per week to two tubes per night. This high dosage produced hallucinations, but when he tried to stop using the drug he experienced uncontrollable tremors. After sniffing six tubes at once, the man slipped into a coma and was admitted to a clinic. Although he recovered, he resumed his habit shortly after being released.

During this period, inhalant abusers tried getting high not only from gases but also from **aerosols** (tiny liquid or solid particles suspended in gas), **propellants** (pressurized gas), and **refrigerants** (substances used to reduce heat or create cold). Some of these substances included cooking sprays, cold-weather engine starters, air sanitizers, window cleaners, deodorants, and even insecticides (chemicals that kill insects)! Although nitrous oxide once again became one of the most widely used inhalants, recent restrictions have made it more difficult to obtain.

By 1993, surveys showed that about one in five eighth graders (19.4 percent) had used some type of inhalant. Inhalant use by all age groups increased during the 1980s.

Who Uses Inhalants?

In the 1990s, inhalants were most commonly abused by adolescents aged 12 to 17. A yearly study called the National Household Survey on Drug Abuse showed that the percentage of people in this age group who first tried inhalants doubled between 1991 and 1996, from 10.3 percent to 21 percent. In the past, more boys abused inhalants than girls, but now the number of females trying this risky behavior is almost as high as the number of males.

Some minority groups, like Native Americans, have

an especially severe problem with inhalant abuse. Sometimes, a fad for inhaling solvents breaks out in a particular school or region of the country. In most places, inhalant abuse has become a greater problem than most adults realize. This is probably because thousands of common products can be abused as inhalants, and most are easily obtainable in drugstores, supermarkets, and hardware and auto supply stores. Inhalant abuse is a problem in other parts of the world as well. Africa, Asia, and Latin America are also struggling to fight inhalant abuse.

One of the most important facts to remember about inhalant abuse is that youngsters who inhale solvents are more likely to try other drugs, like alcohol, tobacco, and marijuana. These three substances are often referred to as **gateway drugs** because young people who experiment with them are more likely to move on to more dangerous drugs (such as heroin and cocaine) than kids who never try the gateway drugs.

There is good news, however. Most kids in the United States have never abused inhalants. If you've never used dangerous drugs like inhalants, you're in good company. It may seem as though most kids are trying drugs, but facts show that the opposite is true. In the next few chapters, we'll look at how inhaling solvents can damage your body and mind and why it's a good idea to stay away from them.

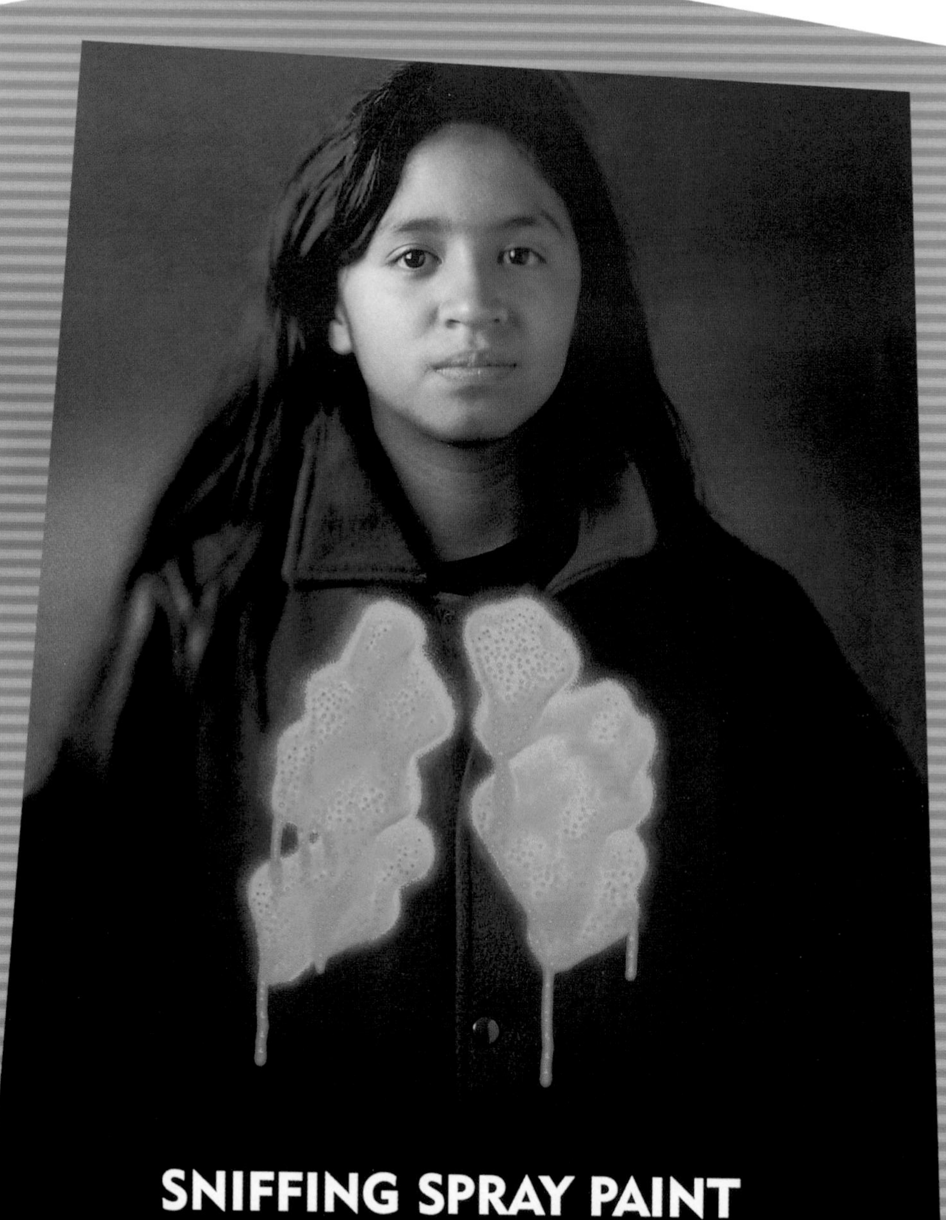

This chapter will tell you how inhalants can damage your lungs—or even kill you.

WHAT INHALANTS DO TO YOU

Some young people are unaware of the harm that inhaling solvents can cause. These youngsters mistakenly assume that because most of these substances are not illegal to purchase and may be easy to obtain, the danger from misusing them is low. Unfortunately, kids who think this way are dead wrong. Volatile organic solvents are among the most harmful substances known. Even worse, the way they are abused—by inhaling—is one of the most deadly ways to take a drug.

According to medical experts, death from inhaling solvents can occur in at least five different ways:

1) **asphyxia**—Inhaling fumes interrupts the abuser's breathing, limiting the amount of oxygen taken in.
2) suffocation—The user becomes unable to breathe.

This is most often seen with abusers who huff substances out of bags.

3) choking on vomit—Because inhalants are toxic to the body, abusing them may cause the user to vomit.
4) careless or dangerous behavior in hazardous settings—Some people, such as certain factory workers, are exposed to solvents on the job and need to take special precautions to avoid being poisoned.
5) cardiac arrest (heart attack)—Some volatile organic solvents can cause the heart to beat irregularly or too quickly. They can also keep the heart muscle from contracting; as a result, the body is not supplied with enough blood.

How Inhaled Solvents Work in the Body

A drug that is injected or swallowed must first pass through the liver, which often **detoxifies** it, or breaks it down into non-toxic substances. Through this process, the liver protects the body from some poisons. A drug that is inhaled, however, travels directly to the lungs and then to the brain. Because an inhaled drug has not been **metabolized** (broken down by the body), it has a much stronger effect on the brain and works much more rapidly. An inhalant takes only about eight seconds to reach the brain!

Once a drug is in the bloodstream, it is absorbed by tissues it encounters. How much damage a solvent does to the body depends in part on its **solubility**, or ability to mix with blood, water, or tissue. Many solvents are **lipophilic** (attracted to fat). That means that they

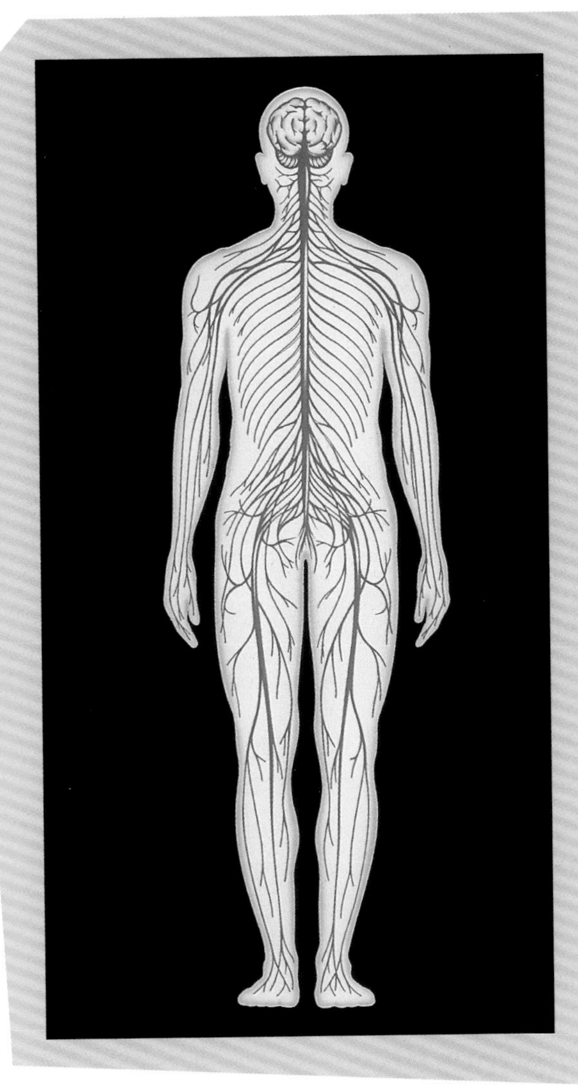

The most important parts of the nervous system, shown in this illustration, are the brain and spinal cord. They act as the control center for the whole body. Abusing inhalants is like dropping a bomb on this control center—sniffing or huffing may damage or even destroy your ability to think, learn, remember, recognize familiar things, and coordinate your body's movement.

become more concentrated in areas that are rich in fat, like the brain and heart. In addition, solvents spread more rapidly into parts of the body that have many blood vessels, such as the lungs, kidneys, brain, and heart. The higher the concentration of the drug in a certain region of the body, the more damage occurs there. Some of these substances also inflame or deform blood

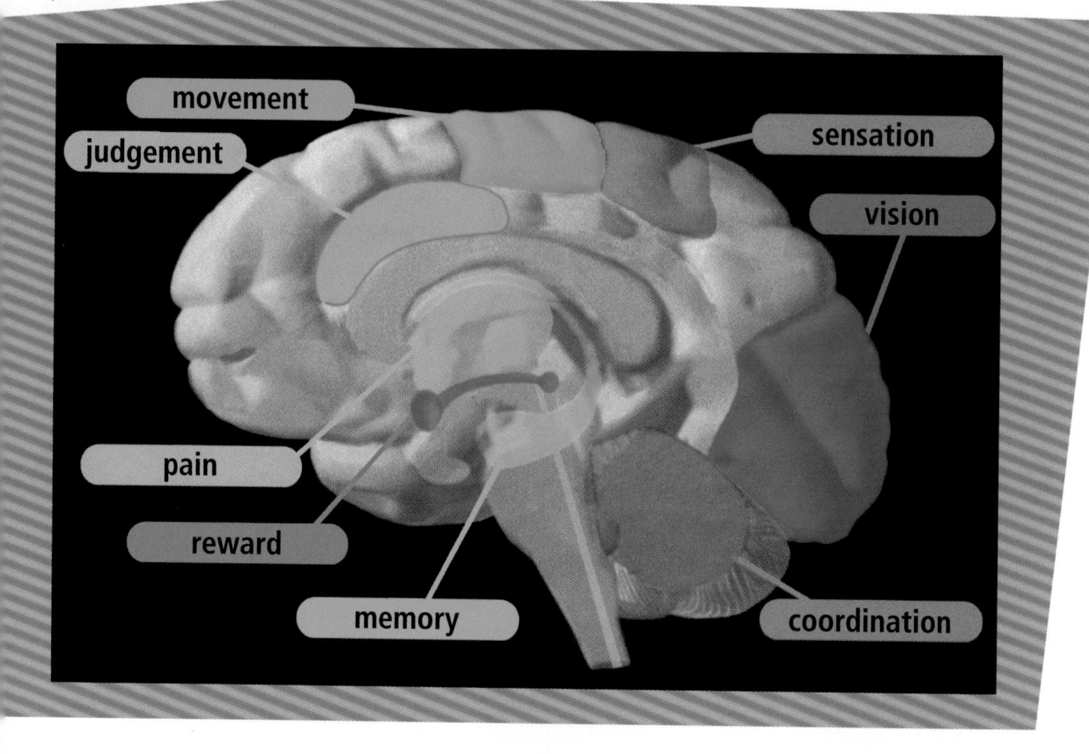

The brain is a very complicated organ. Each of its regions has a special job to do. Substances like inhalants prevent the brain from functioning the way it should. This map shows you where some of the regions are located and what their jobs are.

vessels. They may make the vessels less **permeable** (less able to have materials pass into and out of them).

Inhalants and Your Heart

Inhaled solvents interfere with the normal functions of the heart. In experiments with animals, researchers discovered that heptane and other inhalants, especially propellants, produce an abnormal pumping action of the heart. Certain types of solvents can also disrupt the rhythm of the heart, causing it to beat irregularly. This condition is called **arrhythmia**.

Other inhaled substances, such as butane, a chemical found in cigarette lighters and refills, increase the rate at which the heart beats.

Inhalants can also make the heart more sensitive to a substance called **norepinephrine**, a chemical produced by the adrenal glands and released by the body during stressful situations. Norepinephrine tells the heart to beat faster and raises the blood pressure. The combination of an inhalant and the release of norepinephrine overstimulates the heart and in some cases causes it to stop. Some inhalant users have died this way. The cause of death is known as **sudden sniffing death syndrome**.

Another way that inhalants affect the normal functioning of the heart is by depressing its ability to contract (or squeeze together to pump blood). As a result, the body, including the heart itself, is not supplied with the usual amount of blood. Inhalants also irritate the inner lining of the lungs, and this leads to an increase in heart rate.

Inhalants and Your Brain and Nervous System

In many ways, the brain is the most fascinating and important organ in the body. The brain determines your personality—who you are and what you think, feel, and do. The brain allows people to build skyscrapers, write symphonies, solve algebraic equations, invent airplanes, speak languages, feel love and hate, dream up new ideas, make decisions, and read books. At the same time, the

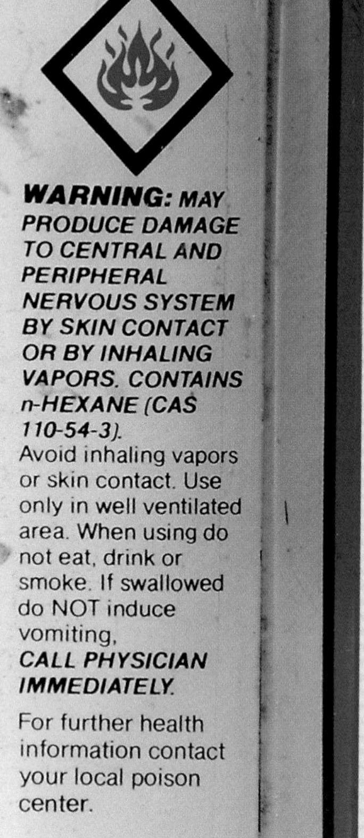

There are probably a number of products in your home or school that carry warning labels similar to the one on this container of rubber cement. You can find more information by contacting your local poison control center.

brain regulates unconscious activity in the body, such as your heart rate, digestion, and breathing.

This complicated and wonderful organ is the headquarters of your body's command and control center, called the nervous system. Elongated nerve cells known as **neurons**, located in the brain and spinal cord, send out and receive messages that govern everything you do. The human brain contains about 100 billion neurons. It is protected by two barriers: a liquid that helps absorb shocks, and the large bone known as the skull.

One of the fatty tissues where inhalant vapors can

concentrate is called **myelin**. Myelin is a white, protective cover that surrounds nerve cells the way the outside coating of an electrical cord protects the wires inside. When inhalants are abused over a long period, they break down the myelin and prevent nerve cells from transmitting messages properly.

Damage from long-term inhalant abuse can slow down or stop nerve cell activity in some regions of the brain. One part that can be injured is the **frontal cortex**, the area of the brain that allows you to think ahead and solve complex problems. This region is also responsible for processing information for the senses.

Another brain region affected by inhalant abuse is the **cerebellum**, which controls movement and coordination. Inhalants can also harm the **hippocampus**, which is responsible for memory and learning. A person whose hippocampus has been damaged will have trouble recognizing familiar things, learning new facts, and following conversations.

Abusing inhalants also changes how neurons generate electrical activity. Normally, neurons are constantly firing and producing low levels of electrical activity in the brain. When this is disrupted, severe brain seizures can result. Changes in the brain's electrical activity can also disrupt the ability to concentrate. Repeated use of inhalants affects electrical activity in the area of the brain that regulates levels of arousal (stimulation or excitement).

Another region of the brain that can be severely affected by exposure to inhalants is the **amygdala**, which regulates emotional behavior. Experiments with animals have shown that repeated exposure to inhalants such as paint thinner can cause **hallucinations** (seeing things that aren't there), **convulsions** (strong and involuntary muscle contractions), lack of coordination, and other problems.

Inhaling solvents can also cause **hemorrhages** (excessive bleeding) in various regions of the brain. Frequently, people who use inhalants develop **insomnia** (trouble sleeping) and may become very depressed. Some cases of alcohol abuse have caused lesions in brain tissue, which may lead to dementia (insanity). Inhalants may act in a similar way.

Inhalants and Your Lungs

One of the first areas damaged by inhalant abuse is the lining of the lungs. Many people don't realize how complex and important the lungs are and how easily they can be injured. Your lungs are not just a set of bags inside you that collect air. They are a vital part of a life-support system that supplies your whole body with oxygen and eliminates waste products like carbon dioxide.

When you inhale, you pull air through your nose and mouth and into the windpipe (also called the **trachea**). On the way to the lungs, the air you breathe is warmed, moistened, and cleansed before reaching the **bronchi** (branches of the trachea).

About halfway down the chest, these branches split and take air into the right and left lungs. The bronchi are lined with special hair-like cells called **cilia**, which

One of the effects of inhalant abuse on the brain is difficulty learning new things. More serious problems, such as hallucinations, depression, and uncontrolled bleeding in the brain, can also develop.

act like tiny filters, constantly waving to help remove dirt from the air. Smoking tobacco or marijuana or using inhalants impairs the cilia and prevents them from cleaning the air that reaches your lungs.

The two main bronchi divide and redivide into smaller air passages that end in tiny air sacs called **alveoli**. The lungs contain millions of these miniature balloon-like pouches. Each lung is about the size of half a football and can hold three or four quarts of air at most (normal, quiet breathing takes in about a pint of air). Because of the alveoli, however, there is so much tissue inside the lungs that if this organ were flattened and spread out, its surface would cover nearly 100 square yards. That's an area larger than a football field!

Each of the alveoli is wrapped in tiny blood vessels that absorb oxygen and eliminate carbon dioxide. Oxygen from the air is removed and passed into the bloodstream by red blood cells. At the same time, carbon dioxide is transported through the bloodstream into the lungs, where it is eliminated when you breathe out (exhale).

When a solvent is inhaled, it passes from the lungs into the blood by way of the alveoli. The drug then goes directly to the brain. Inhaling solvents can irritate the lining of the lungs, increasing the heart rate. It can also cause painful swelling of the lung tissue.

Inhalants and Other Parts of Your Body

Some solvents damage the lining of the liver and cause blood vessels in the kidney to burst. They can also damage bone marrow. This affects the body's ability to produce new blood cells and create antibodies to fight infection. Inhaling solvents can also harm the reproductive organs. If you abuse inhalants when you're young, it may be difficult for you to become a parent when you grow up. In addition, a woman who uses inhalants while pregnant can injure or even kill the baby.

Several solvents may be **carcinogenic** (cancer-causing). Scientists have already discovered that the solvent called benzene can cause or contribute to cancer growth. Other substances that may be carcinogenic are chloroform and formaldehyde.

All of your senses—sight, hearing, touch, taste, and smell—can be damaged by inhalants. Industrial cleaning solvents such as carbon disulfide can cause tunnel vision, a kind of blindness in which only the area directly ahead of the viewer is visible. Drinking methanol (wood alcohol) rather than ethanol often causes total blindness. Toluene, an ingredient formerly used in glue and other industrial products, can destroy high-frequency hearing. Hexane and some ketones may cause the abuser to lose feeling in the hands and feet. In some cases this effect can be reversed if the user stops inhaling these substances, but sometimes the damage is permanent.

By now, you may be wondering why anyone would inhale solvents deliberately. In the next two chapters, we'll look at some of the reasons people turn to drugs like inhalants. We'll also examine some of the ways you and your friends can avoid or stop using them.

Even if you don't want to use drugs, you may feel pressure to do so from your friends. Some ways to avoid situations where alcohol or other drugs may be offered is to find new after-school activities, or organize your own drug-free events like dances, movies, or walk-a-thons.

WHY DO PEOPLE USE INHALANTS?

When youngsters reach adolescence, they begin the sometimes difficult transition from childhood to adulthood. Erik Erikson, the author of a famous book on childhood and society called *Insight and Responsibility*, wrote that one's teen years are a time of feeling uprooted. "Like a trapeze artist, the young person in the middle of vigorous motion must let go of his safe hold on childhood and reach out for a firm grasp on adulthood," Erikson wrote.

Some adolescents don't feel as close to their parents or siblings as they once did. It might feel like friends understand you better because they're going through the same kinds of changes. And while you're trying to figure out exactly who you are, your body has also begun to change or will very soon. These upheavals may make you feel insecure, angry, or frightened. You may find that you want to rebel against your parents and other grown-ups who have authority, such as teachers and religious leaders. Some kids who feel this way mistakenly believe that using drugs and engaging in other types of risky behavior are acts of independence that can make them feel better about themselves.

Using Drugs as a Way to Rebel

If you think about why you experience certain emotions, you will realize that using drugs is not the answer to your problems. Rejecting healthy practices, safety, and common sense just because grown-ups tell you that they are important will not help you. Making an unwise choice, such as inhaling a dangerous substance, shows that the drug user is still a child who needs to be protected by adults.

Peer Pressure

We all know what it feels like to want to belong to a social group. No one wants to feel disliked or left out when a friend is having a party or a group of kids is hanging out. But what happens when someone in the group does something you don't agree with or don't want to do? You want to fit in; you want people to think that you're cool.

This is peer pressure—the feeling we talked about in Chapter 1. Sometimes it's easy to spot. A friend might come right out and say something to make you feel bad if you don't do what everyone else seems to be doing. Other times, your friends might not say anything

The next time you think about trying drugs, ask yourself whether it's really what you want to do. Getting high just to rebel against grown-ups will only create bigger problems.

specific, but you still worry that they won't like you if you don't go along with them.

Peer pressure is one of the main reasons kids begin using drugs. Even adults feel peer pressure, but when you're still growing up and trying to figure out exactly where you fit in, that pressure can be much stronger.

But you don't *have* to give in to peer pressure. There are ways to avoid doing things you don't want to do while still feeling like part of the gang. Chances are, you're not the only one in the group who doesn't want

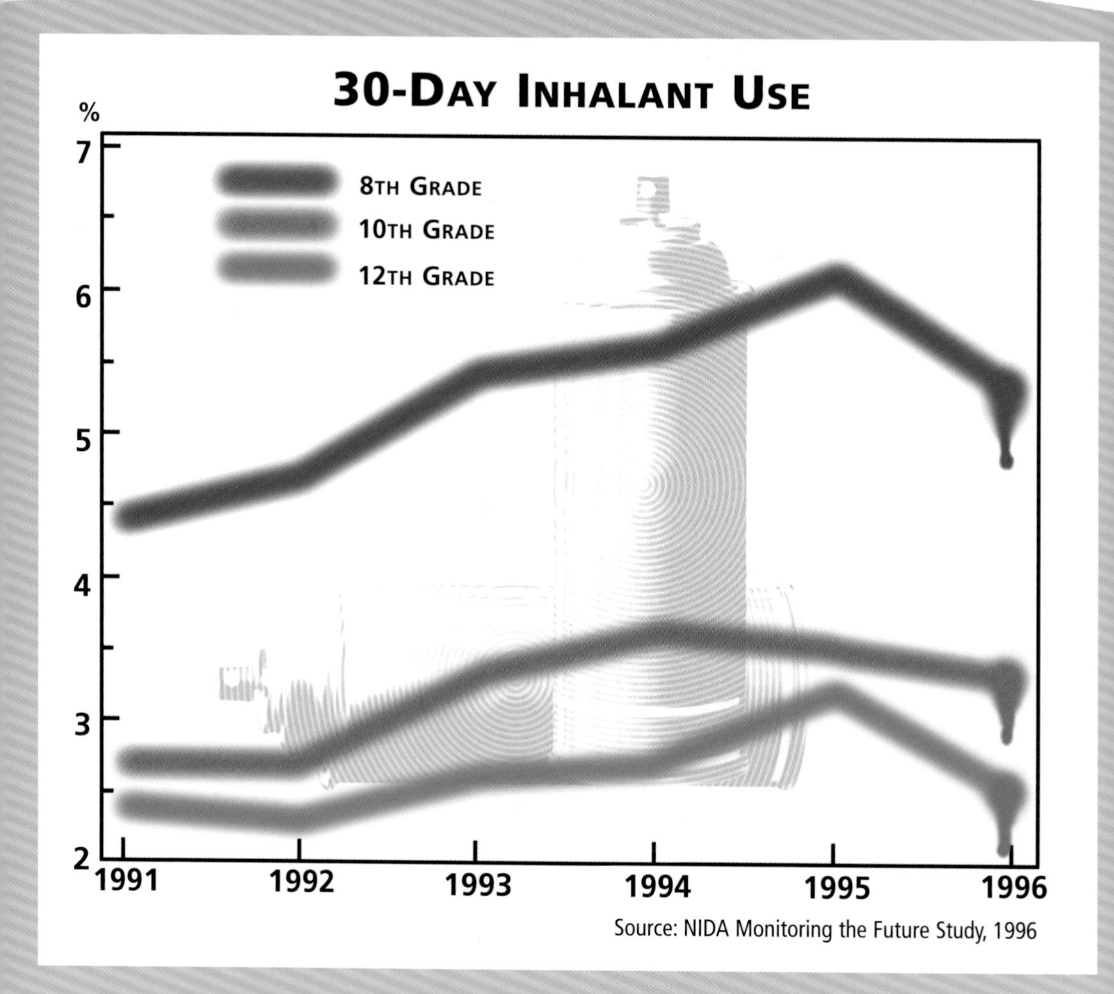

This graph shows the results of six yearly surveys about inhalants by the National Institute on Drug Abuse. Notice that the number of tenth graders and twelfth graders who used inhalants in the month before each survey stayed below 4 percent, but the number of eighth graders who used inhalants rose from about 4.5 percent in 1991 to more than 6 percent in 1995. Except for alcohol and cigarettes, inhalants are the most widely used class of abused drugs among eighth graders.

to use inhalants or other drugs. Other kids may feel the same way, but they may also be afraid to speak up. If you want to do what's best for you, though, you have to stand up for yourself.

This isn't always easy, of course. It takes courage. You may fear that you'll lose your friends. But you may also find that if you do say no, other friends who may have been afraid to speak up will take your side because they'll know they are not alone. And it's often easier than you might think to say "No, thanks."

Here are a few other ways that you can take control of your life and resist drugs:

- Skip parties or activities where you know there will be alcohol or other drugs.
- Hang out with friends who don't drink or use other drugs.
- Find new after-school activities. Check with recreation centers, youth clubs, libraries, or other local organizations to see whether they offer tutoring, sports, music lessons, or craft classes.
- Get involved in drug-free activities like dances, movies, community service projects, or walk-a-thons. Ask your friends to join you.
- Organize a drug awareness program for your school, church, or community center.

Besides feeling peer pressure, here are a few other reasons young people have given for beginning to use drugs:

- To have fun or relax
- To feel more grown up
- Out of curiosity
- Because it seems exciting
- To stop feeling lonely or depressed

All of these reasons reflect perfectly normal emotions. What's important is that you look for ways to satisfy your sense of adventure or curiosity, or help yourself feel less lonely or sad, in ways that are healthier than using drugs.

If anyone asks you to try inhaling a common product, chemical, or unusual substance, tell them you don't breathe poisons or do drugs and that you're not interested in experimenting with chemicals that can hurt you. Be your own best friend by protecting your body.

"Instant" Society

Our society moves a lot faster than it used to. We can contact someone on the other side of the world instantly by using a telephone, fax machine, or Internet connection. Airplanes and even cars carry us at speeds that were unthinkable a few hundred years ago. Appliances such as microwave ovens help speed up processes that once took much longer. Advances in science and medicine help prevent and cure numerous illnesses that were once fatal and can shorten the time needed to recover from sickness, accidents, or disease.

All these improvements have made life easier and better for millions of people. But these changes have

Fax machines, the Internet, supersonic jets, and other modern inventions may offer us speed and convenience. But despite what you may hear from others, drugs do not offer the same "quick fix."

also had another effect. Because everything seems to be moving and working so quickly, some people expect immediate solutions to all of their problems.

Many important parts of life—learning, being creative, growing physically, and accomplishing great things—take time and energy. Overcoming serious difficulties also requires endurance and concentration. Some people become easily frustrated when their desires are not met or their problems cannot be solved right away. A few decide that taking chemicals into their bodies will bring instant happiness, pleasure, or insight. Instead, they get misery, discomfort, more problems, and sometimes addiction.

Mixed Messages

Companies eager to sell their products often use words or images in their advertising to suggest that these items will provide instant gratification (meaning reward or pleasure). If you use this toothpaste, wear this clothing, or drive this car, the ads seem to say, you will be successful, beautiful, popular, and happy.

Some of the models used to advertise products in recent years have been made to look as though they are sick from using drugs. The so-called "heroin chic" style of modeling features pale, underweight young men and women with dark circles under their eyes, dressed in torn or scanty clothing. Seeing these models on billboards, magazines, and television may give some teens the idea that heroin and other drugs are glamorous or cool.

But when you think about it, this is actually a different kind of peer pressure. Advertisers count on your looking at these images and seeing the models as kids like you—and wanting to buy the company's products to fit in. If celebrities and models are using drugs, the ads seem to say, why not you?

You know why. Taking drugs is foolish. It endangers your life. It can destroy your health, your family, and your grades. It is important to remember that *most kids don't take drugs of any kind*. The great majority of young people never abuse solvents, inhalants, or other drugs. If TV shows, movies, magazine ads, radio programs, video games, music, or Internet sites try to idealize drugs, they are doing so with hopes of selling you something.

Use your common sense. Don't sell yourself short by falling for such material. Drugs like inhalants and solvents will never magically provide excitement, sex appeal, ecstasy, or enlightenment any more than the other products advertisers want you to buy.

What if you or your friends have already tried drugs? Is it too late to stop? How can you help someone who is caught up in drug abuse? Chapter 5 will help you find some answers to these questions.

These students in Greenville, South Carolina, are marching to make others more aware of the dangers of drugs. Does your school have a program that teaches drug awareness? If not, maybe you can help start one.

GETTING HELP

Discovering if a friend or loved one has a problem with inhalants and solvents—or with any illicit drug, including alcohol, tobacco, and marijuana—isn't always easy. Children as young as fourth graders sometimes begin abusing volatile organic solvents or experimenting with other drugs. Most inhalant abusers are **polydrug users**, which means that they use many drugs depending on what is available.

How to Tell If a Friend Has a Drug Problem

Most people who are having difficulties with drugs will not simply go to a close friend and ask for help. In fact, drug abusers are more likely to deny the problem and try to hide the symptoms. They may be embarrassed or afraid to confide in someone else. Still, there are some warning signs that you can look for if you suspect that a friend or loved one is abusing alcohol or other drugs. If someone you know displays one or more of the following traits, he or she may have a drug problem:

- Getting high on drugs or getting drunk on a regular basis
- Lying about the amount of alcohol or other drugs being used
- Avoiding you or other friends to get high or drunk
- Giving up activities such as sports, homework, or hanging out with friends who don't drink alcohol or use other drugs
- Having to use increasing amounts of the drug to get the same effect
- Constantly talking about drinking alcohol or using other drugs
- Pressuring other people to drink alcohol or use other drugs
- Believing that alcohol or other drugs are needed to have fun
- Getting into trouble with the law or getting suspended from school for an alcohol- or other drug-related incident
- Taking risks, including sexual risks or driving under the influence of alcohol or other drugs
- Feeling tired, run-down, hopeless, depressed, or even suicidal
- Missing work or school, or performing tasks poorly because of drinking or other drug use

How to Tell If You Have a Drug Problem

Problems with alcohol and other drugs affect all kinds of people, no matter who they are, where they live, or how much money they have. If you abuse alcohol or other drugs and think you're not like others who do the same, you're wrong. Just like everyone else who abuses drugs, you can seriously damage your body and mind, or even end your life. To find out whether you have a drug problem, try to answer the following questions honestly:

- Can I predict the next time I will use drugs or get drunk?
- Do I think that I need alcohol or other drugs to have fun?
- Do I turn to alcohol or other drugs to make myself feel better after an argument or confrontation?
- Do I have to drink more or use more drugs to get the same effect I once felt with a smaller amount?
- Do I drink alcohol or use other drugs when I'm alone?
- When I drink alcohol or use other drugs, do I forget certain segments of time?
- Am I having trouble at work or school because of alcohol or other drug use?
- Do I make promises to others or to myself to stop drinking alcohol or using other drugs, but then break them?
- Do I feel alone, scared, miserable, or depressed?

If you answered "yes" to any of these questions, you may have a drug problem. But don't be discouraged. You're not alone. Millions of people around the world have taken control of their lives and are now drug-free and healthy. Look in the back of this book for suggestions about where to get help for your problem.

Keep in mind that some of these signs, such as mood changes, poor job or school performance, and depression, might be signs of other problems. They could also be symptoms of an illness that you may not know about. Nevertheless, these danger signs indicate that something is wrong.

If someone you know seems to have a drug problem, you might want to talk to him or her to find out for sure. If you do, you may want to follow these suggestions:

- Carefully plan what you want to say and how to say it.
- Pick a quiet and private time to talk.
- Don't try to talk about the problem when the person is drunk or high.
- Use a calm voice, and don't get into an argument.
- Let the person know that you care. Ask whether you can do anything to help, such as finding a counseling or drug abuse treatment center, and offer to go along.
- Don't expect the person to like what you're saying. But stick with it—the more people express concern, the better the chances of your friend or loved one getting help.
- Remember that it's not your job to get the person to stop using drugs. You can offer help, but only they can decide to quit.

Finally, be sure to talk to an adult you trust or one who is trained to recognize drug and alcohol abuse. A doctor, nurse, counselor, scout leader, coach, or parent can give you advice about what to do next.

Effects of Inhalant and Solvent Abuse

Because there are so many different types of inhalants and solvents, the symptoms of abuse are wide-ranging. Using inhalants just once can cause severe mood swings, hallucinations, numbness and tingling in the hands and feet, suffocation, or sudden death.

Short-term effects of inhalant abuse include headaches, dizziness, difficulty breathing, and heart palpitations (irregular heartbeat). In addition to these problems, people who abuse inhalants and solvents over an extended period risk having some or all of the following:

- abdominal pain
- hepatitis (inflammation of the liver)
- involuntary passing of urine or feces
- liver, lung, and kidney impairment
- loss of the sense of smell
- muscle weakness
- nausea
- nervous system damage
- nosebleeds
- permanent brain damage
- violent behavior

When inhalant abuse continues over a long period, the body develops a **tolerance** for these substances. A user will turn to inhalants more frequently and need greater quantities to get the same effect. **Physical**

dependence also can result. A user who tries to stop inhaling solvents may experience **withdrawal** symptoms, including hallucinations, headaches, chills, tremors, and stomach cramps.

Signs of Inhalant and Solvent Abuse

If you suspect someone of abusing inhalants, look for these signs:

- unusual breath odor
- chemical odor on clothing
- dazed, dizzy, or drunk appearance
- signs of paint or similar products on face, fingers, clothing, or elsewhere
- loss of appetite or nausea
- slurred or disoriented speech
- red or runny eyes and nose
- spots or sores around the mouth
- anxiety or irritability
- excitability or restlessness

Some of these symptoms may disappear shortly after inhalation. Other signs that someone has been abusing inhalants and solvents for a long time include malnutrition (the person seems thin and unhealthy) and disregard for personal appearance (the person seems not to care about his or her looks). Inhalant abusers may also carry rags (for absorbing solvents), lots of felt-tip pens, bottles of correction fluid, or other products that produce fumes. Or they may store some of these things in their school lockers or bedrooms.

Use your head—don't sniff or huff dangerous chemicals. A few seconds of feeling high is not worth losing your life.

Inhalants and the Law

In the United States, most drugs are controlled by the Drug Enforcement Administration (DEA), which lists substances under five schedules. Drugs that have the lowest medical value and greatest potential for abuse, such as heroin and LSD, are on Schedule I. Drugs with high medical value and a low possibility for abuse, such as some over-the-counter drugs, are on Schedule V. In between these two extremes are drugs like morphine (for pain control), which is on Schedule II, and some sedatives and stimulants, which are on Schedule III. A doctor's ability to prescribe a medication, and the legal penalties for possessing, using, or selling a substance, depend on where it appears on the DEA schedule.

Most inhalants and solvents are not listed by the DEA because their ingredients are common in household and work products. Toluene, acetone, and octane, for example, can be bought without a prescription or license from industrial suppliers, chemical companies, and even drugstores. These substances are also less expensive than most illegal drugs. Only a few solvents, like amyl nitrite, are dispensed by prescription.

Because many abusable solvents are easy to find or produce, this type of drug abuse is often overlooked by people who study drug addiction and drug control. Parents, teachers, and other adults who work with young people are sometimes unaware of the dangers of inhalant and solvent abuse. But as you now know, these

One of the ways you can help yourself say no to drugs is by exercising. If your body feels strong and healthy, why wreck it with drugs?

substances threaten everyone's health and well-being when used improperly. Now that you have information about inhalants and solvents, you can help make others aware of how deadly they can be.

Getting Help

No matter where you live, you can find help for drug problems from numerous national, state, and local organizations, treatment centers, referral centers, and hotlines throughout the United States and Canada. There are different kinds of treatment services and centers.

Some require their patients to remain at the center as **inpatients** for several weeks. Others provide **outpatient** counseling, meaning that patients attend scheduled therapy sessions but are free to return home after each treatment.

Various resources are listed in the back of this book. Some of the resources you may find in your community are:

- Drug hotlines
- Treatment centers
- City or local health departments
- Local branches of Alcoholics Anonymous, Al-Anon/Alateen, or Narcotics Anonymous
- Hospitals or emergency health clinics

Maybe you are hesitant or fearful about seeking help. It may comfort you to know that most drug treatment programs are designed to provide group (or family) therapy for people with alcohol or drug problems, so you will not have to face your troubles alone. All you have to do is pick up the phone and take the first step. The trained and experienced people on the other end of the line will take it from there.

Some people go through treatment a number of times before they recover completely. Don't give up if you aren't successful right away. Unfortunately, we do not have specific medicines that can cure people suffering from inhalant abuse. Perhaps future research will develop such drugs for use against volatile organic solvents.

Try to keep this in mind, however: research shows

that when a drug abuser gets appropriate treatment—and when he or she follows the prescribed program—*treatment can work.* Getting treatment not only helps users conquer their drug problems, but it also gives them the skills and strength they need to avoid using drugs in the future.

GLOSSARY

aerosol—a substance made of tiny liquid or solid particles suspended in a gas.

alveoli—tiny air sacs in the lungs. A single air sac is called an alveolus.

amygdala—the region of the brain that controls emotional behavior.

anesthetic—a substance that causes loss of feeling or consciousness and therefore helps in relieving pain.

anoxia—a state in which the body suffers from a lack of oxygen.

arrhythmia—an irregular heartbeat that can lead to heart failure.

asphyxia—unconsciousness or death resulting from lack of oxygen.

bronchi—the two main branches of the trachea that carry air to the lungs.

carcinogenic—causing or contributing to the growth of cancer. Many solvents are thought to be carcinogenic.

cardiac arrest—heart attack.

cerebellum—the part of the brain that controls coordination and movement.

chloroform—a heavy, colorless liquid that was once used as an anesthetic. Chloroform was also abused as an inhalant.

cilia—tiny hair-like cells that line the bronchi and move continually to clean air that is breathed in.

coma—an unconscious state from which a person may or may not awaken.

compulsive—a behavior that is very difficult to control. Inhalant abuse is an example of compulsive behavior.

convulsion—a sudden and violent contraction of the muscles.

depressant—a substance, such as alcohol, that slows down the body's central nervous system.

detoxify—to make less poisonous or remove a poison.

ethanol—the type of alcohol found in beer, wine, and hard liquors.

ether—an organic solvent with anesthetic qualities that is used as a medicine to make people unconscious during surgery. Ether can also be abused as an inhalant.

fermentation—the chemical process by which the sugar in a liquid turns into alcohol and a gas. Yeast and certain bacteria cause fermentation in fruit juices.

flammable—easy to set on fire and able to burn rapidly. Many solvents are flammable.

frontal cortex—the part of the brain that thinks ahead and solves complex problems.

gateway drug—a relatively weak drug whose use may lead to experimentation with stronger drugs like cocaine and heroin. Alcohol, tobacco, and marijuana are considered gateway drugs.

hallucination—an object or vision that is not real but is perceived by a person who has a mental disorder or who is using drugs.

hallucinogenic—causing a person to perceive objects or visions that are not real, or distorting a person's perception of objects or events

hemorrhage—heavy or uncontrollable bleeding.

hippocampus—the part of the brain responsible for processing memory.

huff—to breathe in an inhalant or solvent through the mouth.

inpatient—a patient who stays at, or is checked into, the clinic or hospital where treatment takes place.

insomnia—difficulty sleeping or an inability to sleep.

lipophilic—attracted to areas or substances that are rich in fat (lipids).

membrane—a thin, flexible layer of tissue in the body of an animal or plant that usually lines or covers part of the body.

metabolize—to break down food into energy and living tissue and then dispose of waste materials. All living things metabolize.

myelin—the protective cover that surrounds the body's nerve cells.

neuron—a nerve cell in the brain or spinal cord.

norepinephrine—a chemical released by the body during stressful situations that causes the heart to beat more rapidly.

organic—containing (or derived from substances that contain) carbon and hydrogen.

outpatient—a patient who visits a hospital, clinic, or other facility for treatment but is not required to check in.

peer pressure—words or actions by friends, siblings, or someone else the same age that make a person feel as though he or she has to act like them to fit in with the group.

permeable—allowing substances to pass or spread through.

physical dependence—a state in which a drug user's body chemistry has adapted to require regular doses of the drug to function normally. Stopping the drug causes withdrawal symptoms.

polydrug user—a person who uses a number of different drugs. Most inhalant abusers are polydrug users.

propellant—a pressurized gas used in bottled or canned products such as hair spray and whipped cream that helps to expel the contents. Some inhalants are propellants.

psychoactive—affecting the mind or behavior.

refrigerant—a substance used to reduce heat or create cold, as in a refrigerator or air conditioner. Refrigerants are sometimes abused as inhalants.

sniff—to breathe in an inhalant or solvent through the nose.

solubility—the ability of a substance to be dissolved in a liquid or tissue.

solvent—a liquid, such as water, alcohol, or ether, that can dissolve another substance to form a solution.

sudden sniffing death syndrome—a condition in which an inhalant abuser's heart becomes overstimulated by the inhalant and the body's release of norepinephrine, causing immediate death.

tolerance—the body's ability to endure or become less responsive to a drug. A user needs increasingly large amounts of the drug to achieve the same level of "high."

toluene—a volatile organic solvent used in many industrial products, such as paint thinner. Toluene is often abused as an inhalant.

toxic—poisonous.

trachea—the windpipe; a tube that goes from the throat to the lungs and is used in breathing.

tremors—uncontrollable shaking.

volatile—characterized by the tendency to change rapidly from a liquid to a vapor. Volatile substances are usually flammable.

volatile organic solvents—a group of chemicals that can pass readily through the body and damage the brain, heart, lungs, and other organs.

withdrawal—a process that occurs when a person who is physically dependent on a drug stops taking the drug.

Bibliography

Glowa, John. *Inhalants: The Toxic Fumes*. New York: Chelsea House Publishers, 1992.

Health Edco WRS Group. *Inhalants: The Quick, Deadly High*. Waco, TX, 1998.

Inhalants. Syndistar, Inc. 20 min. 1994. Videocassette.

Inhalants: Are You Out of Your Mind? Syndistar, Inc. 15 min. 1997. Videocassette.

Inhalants: Sniffing Your Way to Addiction. Altschul Group Educational Media, Visions Video Production, Inc. 14 min. 1997. Videocassette.

Life Skills Education. *Inhalants*. Weymouth, MA, 1997.

National Clearinghouse for Alcohol and Drug Information. *Straight Facts About Drugs and Alcohol*. Rockville, MD, 1998.

National Institute on Drug Abuse and National Institutes of Health. *Mind over Matter: The Brain's Response to Inhalants*. Rockville, MD: National Clearinghouse for Alcohol and Drug Information, 1997.

A Wasted Breath: Kids on Inhalants. Media Projects, Inc. 19 min. 1992. Videocassette.

Wisconsin Clearinghouse. *Inhalants: Mind-Altering Drugs Series*. Madison, WI, 1998.

Find Out More About Inhalants, Solvents, and Drug Abuse

The following list includes agencies, organizations, and websites that provide information about inhalants and other drugs. You can also find out where to go for help with a drug problem.

Many national organizations have local chapters listed in your phone directory. Look under "Drug Abuse and Addiction" to find resources in your area.

Agencies and Organizations in the United States

American Council for Drug Education
164 West 74th Street
New York, NY 10023
212-758-8060
800-488-DRUG (3784)
http://www.acde.org/
wlittlefield@phoenixhouse.org

Center for Substance Abuse Treatment: Information and Treatment Referral Hotline
11426-28 Rockville Pike, Suite 410
Rockville, MD 20852
800-662-HELP (4357)

Chemical Specialties Manufacturers Association Inhalant Abuse Education Program
1913 I Street, N.W.
Washington, DC 20006

Drugs Anonymous
P.O. Box 473
Ansonia Station, NY 10023
212-874-0700

Eden Children's Project
1035 Franklin Avenue East
Minneapolis, MN 55404
612-874-9441

Families Anonymous
P.O. Box 3475
Culver City, CA 90231-3475
310-313-5800
800-736-9805

Girl Power!
U.S. Department of Health and Human Services
Office on Women's Health
11426 Rockville Pike, Suite 100
Rockville, MD 20852
800-729-6686
http://www.health.org/gpower
gpower@health.org

International Institute for Inhalant Abuse
799 East Hampden Avenue, Suite 500
Englewood, CO 80110
800-231-5165

Just Say No International
2000 Franklin Street, Suite 400
Oakland, CA 94612
800-258-2766

Marin Institute for the Prevention of Alcohol and Other Drug Problems
24 Belvedere Street
San Rafael, CA 94901
415-456-5692

Narcotics Anonymous
P.O. Box 9999
Van Nuys, CA 91409
818-780-3951

National Center on Addiction and Substance Abuse at Columbia University
152 West 57th Street, 12th Floor
New York, NY 10019-3310
212-841-5200
212-956-8020
http://www.casacolumbia.org/home.htm

National Clearinghouse for Alcohol and Drug Information (NCADI)
P.O. Box 2345
Rockville, MD 20847-2345
800-729-6686
800-487-4889 TDD
800-HI-WALLY (449-2559, Children's Line)
http://www.health.org/

National Council on Alcoholism and Drug Dependence (NCADD)
12 West 21st Street, 7th Floor
New York, NY 10010
800-622-2255

National Families in Action
2296 Henderson Mill Road, Suite 300
Atlanta, GA 30345
404-934-6364

National Family Partnership
1159B South Towne Square
St. Louis, MO 63123
314-845-1933

National Inhalant Prevention Coalition
1201 West Sixth Street, Suite C-200
Austin, TX 78703
800-269-4237
512-480-8953
http://www.inhalants.org
nipc@io.com

Office of National Drug Control Policy
750 17th Street, N.W., 8th Floor
Washington, DC 20503
http://www.whitehousedrugpolicy.gov/
ondcp@ncjrs.org
888-395-NDCP (6327)

Parents' Resource Institute for Education (PRIDE)
3610 DeKalb Technology Parkway, Suite 105
Atlanta, GA 30340
770-458-9900
http://www.prideusa.org/

Solvent Abuse Foundation for Education (SAFE)
750 17th Street, N.W.
Washington, DC 20006
202-332-7233

Agencies and Organizations in Canada

Addictions Foundation of Manitoba
1031 Portage Avenue
Winnipeg, Manitoba R3G 0R8
204-944-6277
http://www.mbnet.mb.ca/crm/health/afm.html

Addiction Research Foundation (ARF)
33 Russell Street
Toronto, Ontario M5S 2S1
416-595-6100
800-463-6273 in Ontario

Alberta Alcohol and Drug Abuse Commission
10909 Jasper Avenue, 6th Floor
Edmonton, Alberta T5J 3M9
http://www.gov.ab.ca/aadac/

British Columbia Prevention Resource Centre
96 East Broadway, Suite 211
Vancouver, British Columbia V5T 1V6
604-874-8452
Fax: 604-874-9348
800-663-1880 in British Columbia

Canadian Centre on Substance Abuse
75 Albert Street, Suite 300
Ottawa, Ontario K1P 5E7
613-235-4048
Fax: 613-235-8101
http://www.ccsa.ca/

Ontario Healthy Communities Central Office
180 Dundas Street West, Suite 1900
Toronto, Ontario M5G 1Z8
416-408-4841
Fax: 416-408-4843
http://www.opc.on.ca/ohcc/

Saskatchewan Health Resource Centre
T.C. Douglas Building
3475 Albert Street
Regina, Saskatchewan S4S 6X6
306-787-3090
Fax: 306-787-3823

Websites

Avery Smartcat's Facts & Research on Children Facing Drugs
http://www.averysmartcat.com/druginfo.htm

Center for Alcohol & Addiction Studies (CAAS)
http://www.caas.brown.edu/

Consumer Product Safety Commission
http://www.cpsc.gov

D.A.R.E. (Drug Abuse Resistance Education) for Kids
http://www.dare-america.com/index2.htm

Drug Strategy Institute
http://www2.druginfo.org/orgs/dsi/

Elks Drug Awareness Resource Center
http://www.elks.org/drugs/

Join Together Online (Substance Abuse)
http://www.jointogether.org/sa/

National Institute on Drug Abuse (NIDA)
http://www.nida.nih.gov/

Partnership For A Drug-Free America
http://www.drugfreeamerica.org/

Reality Check
http://www.health.org/reality/

Safe & Drug-Free Schools Program
http://inet.ed.gov/offices/OESE/SDFS

Teen Challenge World Wide Network
http://www.teenchallenge.com

U.S. Department of Justice Kids' Page
http://www.usdoj.gov/kidspage/

INDEX

Picture Credits

page
6: Courtesy Office of National Drug Control Policy; the White House
12: Reuters/Erik De Castro/ Archive Photos
15: Michael Donne/Photo Researchers, Inc.
18: photo © Judy L. Hasday
21: photo © Judy L. Hasday
24: National Museum of Art, Washington DC/Art Resource, NY
26: Archive Photo
28: Graph illustration/Keith Trego
31: photo © Judy L. Hasday
34: National Inhalant Prevention Coalition
37: John Bavosi/Photo Researchers, Inc.
38: Courtesy National Institute on Drug Abuse
40: photo © Judy L. Hasday
43: PhotoDisc Vol. 24 #24036
46: Richard Nowitz/Photo Researchers, Inc.
49: PhotoDisc Vol. 24 #24046
50: Graph illustration/Keith Trego
53: PhotoDisc Vol. 24 #24136
56: AP/Wide World Photos
63: National Inhalant Prevention Coalition
65: PhotoDisc Vol. 45 #45162

LINDA BAYER has an Ed.D. from the Graduate School of Education at Harvard University and a Ph.D. in humanities. Dr. Bayer has worked with patients suffering from substance abuse and other problems at Judge Baker Guidance Center and within the Boston public school system. She served on the faculties of several universities, including the Hebrew University in Israel, where she occupied the Sam and Ayala Zacks Chair and was twice a writer in residence in Jerusalem.

Dr. Bayer was also a newspaper editor and syndicated columnist, winning a Simon Rockower Award for excellence in journalism. She is the author of hundreds of articles, and is working on a fourth book. She is currently a speechwriter and strategic analyst at the White House, where she has written for a number of public figures, including General Colin Powell and President Bill Clinton.

Dr. Bayer is the mother of two children, Lev and Ilana.

BARRY R. McCAFFREY is director of the Office of National Drug Control Policy (ONDCP) at the White House and a member of President Bill Clinton's cabinet. Before taking this job, General McCaffrey was an officer in the U.S. Army. He led the famous "left hook" maneuver of Operation Desert Storm that helped the United States win the Persian Gulf War.

STEVEN L. JAFFE, M.D., received his psychiatry training at Harvard University and the Massachusetts Mental Health Center and his child psychiatry training at Emory University. He has been editor of the *Newsletter of the American Academy of Child and Adolescent Psychiatry* and chairman of the Continuing Education Committee of the Georgia Psychiatric Physicians' Association. Dr. Jaffe is professor of child and adolescent psychiatry at Emory University. He is also clinical professor of psychiatry at Morehouse School of Medicine, and the director of Adolescent Substance Abuse Programs at Charter Peachford Hospital in Atlanta, Georgia.